Science And Religion Are Wrong

Real Facts About The Universe, Evolution, The Paranormal, And Much More...

By: Wise Wolf

ISBN: 978-1-304-80319-1

About The Author

Ever since I was a child, always had the need to understand my paranormal experience and reality it self. Although I grew up in a Christian family, I internally could not except. It was as if my intuition was guiding me to seek else where, in nature and science. Although I loved science as a kid, I wasn't fully open to accept everything because my intuition was telling me science it self can't explain how existence came into being, nor was open to any cosmic truth that scientific minds can every comprehend. Not even New Age knows that I came to know through my own personal self-exploring for both the cosmic and divine truth in which the ego conscious of believer will ever be open to such possibilities. Instead they would rather have me accept their beliefs, not explore for the truth for my self. So I guess you can say I bit stubborn, because what one believes does not mean that is the absolute truth. And it is this internal inspiration that lead me on the path of wisdom and the absolute truth of that in which I came to know and share in my teachings.

About The Book

Unlike science and religion, I do not ask you to accept anything here as the truth. Doing so is giving into belief, and belief is not truth. Instead, I ask you to be your own teacher, your own student when it comes to being a truth seeker. That means you are asked to lose your ego, for it is closed to any truth that doesn't fit into it's belief. Here, your are asked to follow the ways of the ancient sages like The Buddha and those who went into isolation in order to cultivate clarity and wisdom for themselves. This means you must be completely open to any possible truths, which you must question everything. By questioning everything means question your own personal knowledge(you beliefs or theories), question the knowledge of science and religion, other people beliefs, how nature works, and so forth. What you come to know, you should not fully accept. But question it again and again until you have full clarity to that in which you seek clarity on. You are also asked to explore for the truth, rather just accept the possibilities. This means if you truly want to know if there is some cosmic god, ghosts, another reality beside our own, even your own existence? Then you have to explore for it for your self, not accept what others believe or tell as a fact. Instead, you seek for the facts for your self. Never accept it can't be done, doing so robs you from discovering the truth in which you been seeking. Never accept other people's truth, in books, what you see on television, in class, and so forth. For people can be very deceiving, and want you to believe their truth(which may not be true). No matter how much they want you to accept their truth, you have your own free-will and therefore no one can't tell you what to believe. You are your own master of your life, your own teacher, your own leader, your own explorer. But you have to have a an open mind to any possible truth, and you must seek it with no ignorance. Be open in mind, be open to what the cosmos and nature have to share with you. For what you seek, you can come to know it.

Table Of Contents

Chapter One

The Path To Truth

Facts From Fiction

Both science and religion would like us to accept their knowledge are facts, but we come to know that much of their knowledge makes no sense to our curious minds. This is because much of their so-called truth are in fact made-up and unproven. Since they tell us this is their truth, our internal instincts maybe telling us another story? And it this internal intuition we should listen to, because it want's us to know the truth and not the false. This means we must ask those unanswered questions, and be very skeptical to the so-called truth of others. Other wise you may be accepting something false, even denying possible truth. And it is this ignorance is why atheist and mystic believers are always at each other throats about, nor can't respect one another beliefs. As truth seekers, we can't allow ourselves to be caught up between these two belief systems. Truth seekers are a lone, are independent from the social order of ego minded people that tells us what to believe and how to live against our free-will. And it is these social control orders that we should separate ourselves from. Other wise, you rob your self from the absolute and allow yourself to be brain-wash by these egotist social orders.

Question Everything

Science and religion claim they want you to ask a lot of questions about that you need clarity on, but what they are not saying is that they want you to question knowledge that is not about of their truth but that outside it. We experience this all the time. Scientist want us to question religion and see if it all makes any sense. As it is the same with religious people. But they really don't want us to question their truth, rather we must accept it because it is the absolute. That is where you should be skeptical, that is where you should start asking a lot of questions, and not be so willing to accept nor deny what they claim to be the facts. Facts are facts, and facts can be proven true or false. Fiction is fiction, we know what is fiction because our matrix conscious tells us this not the truth but make-believe. It would be wise to listen to your higher conscious, because it holds all the wisdom of the cosmos and nature within its self. The ego conscious doesn't, it only claims to know and make up false truth all in order to gain attention. It is refer to in my early books as the "devil conscious, and the higher conscious as angelic or light conscious. For the ego is deceiving, but the matrix conscious is honest. Don't let your self be deceive by your own ego, as well the ego of others. Follow your internal instincts, your true guide within your matrix conscious that wants you to know what is real and what isn't. For the ego knows no real, it knows only deception. And it is the ego that teaches it's knowledge in science and religion. As well in history and story telling. The higher conscious does not, it reveals the truth though experience and meditating on that we need clarity on.

You Are Your Own Follower

Don't let your self become a slave to our society social order of expectations and what we should believe in. You have free-will, you are your own master of destiny and path to truth seeking. No one owns you, so don't allow society control you. Don't let your mind be brain-washed with deception, allow your higher conscious mind cut through the falseness in order to discover the truth hidden within it. Be your own teach, your own leader, Follow your own instincts, your own free-spirit life. Be you and not what others nor society wants you to be. Follow no one, only follow your own calling and internal inspirations. Always be good, and do good for others. And know only that in which you have full clarity on and experienced, not anything that seems false and might be deceiving. Listen to your higher conscious, it is inspiring your to the true path to wisdom and self-exploring spiritual living.

All Is An Deceiving

If it is truth you seek, never trust what you read nor perceive. For we live in a world full of deception, a society full of lies. This is exactly what The Buddha warns us about, and teaches us how to cultivate clarity and come to know the real. He tells us not accept anything what he says nor what others tell us, what religion and history want us to accept, and not allow ourselves be deceive by what we experience. If it truth we seek, then we must be skeptical of what we accept as truth, and the so-called truth of others. It is the same of what one reads and learns in this book. I ask never accept my teachings as a fact nor as a fiction, because you and you a lone must find clarity and any truth to what you learn here. Other wise you accept or deny that in which that doesn't fit your own beliefs. Belief is a delusion, it is accepting that in which you have no real clarity about, no real experience that lead you to discover that facts in which you keep seeking until you have full clarity about. Don't let your self be deceived by your ego, listen to your higher conscious that knows the absolute.

Knowledge Has Two Sides

knowledge has two sides, the positive and the negative, the real and unreal, the true and the false. Knowing this can help you find the truth you seek, and one way is by not trusting what we learn by the knowledge of others. For what we are told in religion, science, and history had the real truth. Truth that lies beneath the deception of false facts, fiction, and lies. Science would like us to accept their truth is the only truth, what makes atheism and religion indifferent. For both want us accept something that we have not discover the facts for ourselves. Bother believe in something that they don't know nor unsure of. Both believe that fits their ego needs, not fully open to the possible truth in which only the keen minds can easily seek if one ignores their ego and follow their internal intuition. If it is truth you seek, then you have to seek it for your self. No one can give you truth, for the majority do not know the real. Only you and you a lone have to discover that for your self. And that is what I inspiring you to do in this book. Therefore be open in mind, and meditate on everything for greater clarity.

Book Knowledge Is Not Wisdom

Wisdom is knowledge that one fully understood and have years of experience in which gave one greater clarity to that is most truth. Books only teach knowledge, it can not give you wisdom. Wisdom comes with time and personal progress. The more you experience, the more and more you come to understand it. The more you ponder on something, the closer and closer you come to understand it with greater clarity. But accepting that in which you have no full clarity one nor ever experience is cultivating wisdom. Reading this book does not make you wise, having a fuller understand and experience what I experience does. Accepting what you learn in this book does not give you truth, you have to discover the truth for your self. This book only lays out the facts and inspire you to be your own truth seeker. It can't give you truth, accepting without exploring is giving into belief. Only science and religion want you to accept, and give you false experiences in which your own perception manifest.

Do The Research

Before accepting something, it is wise to explore much farther. Find out any historic discoveries that can help give you more understanding of how science and religion came into being, and how no side accept the same belief. Know that both tell a good story, doesn't mean there may not be some truth in the teachings? It also means they may not even know the truth, but fill in the blinks with false truth. Know that truth lies within that in which can not be seen, You have to explore and uncover it. If you want to know if there is some cosmic god? Then you have to find how you can experience it for such possibility. If you want to know if ghosts exists, then you have go out and seek them out. Which I really would not advise, for such can be dangerous. If you want to know how life began, know that at present science are not fully sure, nor is religion. You have to come to realization to such things in which you would have to have been a witness to such events in order to know the truth. And since we can not, we only have theories. Which seems to become beliefs by many. That is what makes atheism not any different from religion. Know wise man never tells anything he is not sure of, nor teach anything that people will not be truly open to. You will become the outcast, the lone wolf of wisdom and truth seeking. Only those who truly want to know will be more open to it. You reading this book is telling your self that you may be open to such possibilities, even want to discover it for your self. And this is the true path to wisdom and truth. This is what this books guides one to seek.

Chapter Two

Creationism

When it comes to how everything came into being, both science and religion have their own creationism. All want us to just accept their truth, but the fact is neither own do not really know. Science would have us believe their truth is truth and religion truth is false. Religion would like us to believe their truth is truth and science truth is wrong. Fact is neither is right, nor neither one is wrong. For the absolute lies in the middle, but the ego mind can see that truth. For to find out what is the absolute, you would actually had to be there at that point in time of the event to be a witness to creation. And since we can not, then how can any one say they know? And it is this clarity is what you should be seeking, not the false truth of both sides of these beliefs systems.

Science would like us to accept that in the beginning there was noting, and out of nothing there was something, and that something was the Big Bang. A theory that was well accepted do to the expansion of the universe, but that may just be an illusion it self? But how is scientist certain? They are not, it has always been a theory that majority come to accept as truth. But there truth only has become a belief, a religion. And as truth seekers, this belief makes no sense, and have so many unanswered questions that scientist may be over looking? Such questions we should be asking ourselves is how can something come out of nothing? How can there be nothing and then something exist? How can something as tiny as a golf ball come into existence from nothingness? How can something come out from this tiny matter and become something greater? How did the cosmos even know how to create what we come to call existence? The only way to find out the real answers is by meditating on them for greater clarity, to make real sense out of them. Then once we have come to understand, we should keep questioning our accepted knowledge for any new information or any false? You may come to know that you will come to gain much more clarity, much more closer to the absolute.

Religion such has Judaism like us to accept in the beginning was nothing except God, and it was God that created everything in certain amount of days. As we come to know from science that is is impossible and could have never happen. The fact of God being real does not fall into their ego beliefs. So all this is sweep to side and hope we throw it all out and accept what science tells us. While religion would like us to accept these stories as truth, be we have came to know these stories make no sense, and that historical evidence tells us how much these stories in Judaism kept changing. Which we come to ask ourselves similar question as we did with the Big Bang theory. If nothing existed, then how did God came into being? If God does exist, why was God existing in empty space for such a long time and then decides to create the universe? Where did God get all these materials? As well many other questions you may be asking. We now know that the universe wasn't created in few minutes nor days. But there is still the question of was there even ever a beginning? Fact is not all religion have a beginning belief, there are some religion tell us existence has always existed. And it is this belief that makes it more of a fact, not fiction. Not just because it makes sense, but it fills in the falseness of something existing out of nothing. That doesn't mean the Big Bang never happen, nor God Never created everything. There has to be be truth the lies between these two beliefs. And maybe that truth is both are right? That the truth only you have to discover for one self. Science and religion can't give you the truth. Nor should you true their truth as truth.

One thing is certain, you can't have nothing without something, nor something without nothing. Before there was something, there had to be nothing, but something had to exist in the first place? Other wise how can something exist from nothing, it had to always existed? And if our universe was created from nothing, then it would make much more sense that the universe existed before it existed out of nothing? What is certain, the cosmos follows the Law Of Change, which part of that law is everything follows a cycle. A cycle of no beginning nor end, that in which was created becomes undone only to be created again. One way to finding out this fact by observation of Nature. We know that Nature follows a cycle of change, it created recycling and keep changing. What is created so meets its end, but it is never gone only take another form. It is the same with the cosmic order. So existence truly has always existed, which what makes present day and the old belief of creationism to be more of a good story. And we know good stories are made up from the imagination of the human ego conscious mind.

Chapter Three

Materialism

The world as we know it seems to be something physical? Something you can sense with your senses and experience as real thing. Science tells us all is made up noting more than energy and matter. Which that was well known by ancient sages of Buddhism and Taoism, but not in such a way as old scientist believed. While some of the ancient religions were you find the belief and practice of Shamanism go much deeper. Many shamans tell of our world is made up of energy, but the energy particles are not lifeless. Meaning energy is a spirit, and all things are made up of these spirits. So who is right? Science or religion? Maybe the answer is both?

Over the decades, physics has been evolving with never ending questions and theories. Today, physics is broken up into two groups." Materialism" and "Quantum Physics". It is the old physicists beliefs everything is made up of nothing but matter, matter which is made up of atoms. Over time the existence of an atom was broken down into many other smaller particles, and at present wonder how small can they go? Then string theory came to be in the ending of the 20th century, but no proof of these forms of energy exist. But that doesn't mean theory will not become a belief, this is expected by the common people. It's part of their primitive nature to believe in something that they are not sure exist or not. But quantum physics goes beyond the physical to a more much more complex theory in which the old physic scientist can't fully comprehend nor accept any possibility. The reasons for this is because their egos are not open to new possibilities, it only wants to accept what fits into it's beliefs. And this is exactly we find atheist do, which they turn their own atheism into a religion. As for quantum physics? There are so many theories of how all came into being, how many dimensions they might be, even our reality being a holographic illusions of energy and lights. So the questions one might ask is who is right? The only real answers is both are. Know that knowledge has two opposite nature, is is the same with everything. That means reality is made up of the physical and nonphysical, the seen and the non-seen realities. Although many scientist are not open to such possibles, but the fact is you can;t have one without the other. And as for a holographic like existence? Fact is matter like a solid is a mere illusion, manifestation of it's solidness created by our own ego perception. In other worlds, the world outside us is an illusion, perceived almost like a hologram. But is there any facts support this unseen reality existing? Problem is common people can't, for they can't see beyond their physical perception. Only those with heighten senses can see through solid as if they have x-rays visions. Which is just reality unfolding to them.

Religion on the other hand depends? When it comes to the mysticals, we find reality is either made up of magic(energy) or the energy from the gods. At present, what is refer to as "New Physics" would more likely agree both is true(in theory anyway). But god in quantum physio is way more different from those made-up in religion. One theory is existence is made up like a hologram, at the center is a matrix in which we perceive through our sense of a presence of a God. In other words, existence is like one big cosmic computer program, and the matrix is God. In Shamanism, you will find a similar belief of some kind of holographic reality that is perceive as dream-like. So who is more right? Fact they are not indifferent, only perceived two different description. For that fact perception is holographic-like in nature, dream-like images or sensations created by countless particles of energy. But one question doesn't seem to be asked? That question is how does energy know how to create things that we can perceive? It is much easier for mystical believers to fill in the blanks that God is behind it all, or it is all does to magical forces at work. Which leads to another unanswered question, is our entire existence have a mind of it's own? Maybe that is the case? It may help best explain how prayers seem to get answered, things that you seek answers about also seem to be answered by what you experience outside your mind and body at come point of event in the week or so? Maybe the cosmos is one big mind, the same mind that created your conscious? Thing is science don't have the tools to prove it, and religious people may not even know the real answer to that question because it may not be part of their religious beliefs? But the fact is, you actually can find out if it is all one big illusion and reality is made up of consciousness? The only problem is you have to be open to the possible that meditation can in fact tune you out of the illusion of the physical perception world and tune into your own higher conscious that is the matrix of your mind. That matrix is linked to the cosmic web, to the cosmic conscious. You just have to practice and find out. Other wise you may truly never know. It's not impossible to experience for your self, but you not going to experience beyond this physical realm because the illusion hides the real reality that lies beyond the physical existence.

At present, science is leading more and more towards reality being more like a hologram, but not some alien virtual reality as some extra terrestrial believer believe. Think of it like clear glass of energy, except it's more fabric-like. Like a web, it is made up of many of these particles of energy. But not all is of form, for there is also the opposite of the formless. Which what Taoist refer to all made up of the Yin and Yang. Which is opposite forces work as one whole, and this is what existence is made up of. Our physical world is what it seems, it just a mere illusion of collective minds of living beings dreaming in heir heads which they believe what they experience outside their bodies is actually real. It is what lies behind the illusion is what is real, and there are only two ways of finding out. Either in death or an out-of-body experience. But not all ignorant minds of many science and religious people are willing to accept. To them, our ghostly self doesn't exist. How funny that will be when they after death as they move about beyond their dead physical shell. It is also funny how many think their spirit or conscious is one big entity, not manifestations of energy particles that seem to have a mind of their own? Maybe because delusional minds accept belief and not truth? And our world is filled with believer, both atheists and religious. Which that's okay, not everybody will know the truth until they reach their finely rebirth beyond their holographic illusion of the many realms of reality. Until then, believers will be believers. We need to respect that, it's not our place to force people to accept what we come to accept as our personal truth. Just as long their beliefs are harmless to all living beings and the environment of our world in which all life forms made home at those locations.

Chapter Four

Other Dimensions

It has long been the belief of humanity of the existence of other realms beyond this one. The many religions seem to always tell of tree or more realms of existence. The most common is the heavenly realm, the earthly realm, and the under world. Then there is the land of the living and the afterlife. Quantum physics have many different theories what these dimensions may look like. Problem is common people are limited to their conscious experience, many don't have the right state of mind to be able to perceive beyond their brain. We only find this higher conscious state of mind in which the original humans have. That is because the original humans had to use their higher perception for survival purposes. This why they are able to perceive beyond not just their surroundings like predatory animals do, but also any threats that hide behind the physical layer in which many believes refer to after life. Something the common people might not experience because they have weaker perceptions in which has limited range and effectiveness. This is do to living their tribal way of life to build and live in cities, which they created new religions that best fit their culture and even political influences of the religious leaders. So living in cities disconnected them form that natural world in which they had to use their higher perception from hidden dangers. But even our cities and communities have human predators, but that doesn't mean these lower perceptions can't still serve as survival instincts. This just means they are weak, and weak perceptions can't sense beyond it's physical surroundings. This is one of the reasons why typical people don't truly know what lies beyond the physical. Why those that can, common people always perceive them as highly delusional. But maybe it is them who are? For they see this dimension as something stable.

Depending one on the religion? You will find many similar stories of other dimensions, as to how they came to be has many possibles. It is a fact that many religious beliefs are just stories that hidden some spiritual teachings, even political control of their people's minds. While the other source may have came from dreams or hallucinations. The other possible source is few of these people actually experience the other dimensions beyond their physical body. Which is the case in near-death and out-of-body exploration. Which tells these dimensions are actually layered and dream like. Others tell of reality that change with their thoughts, even dreamed up ghostly beings into existence which is like a holographic projection. This is the fact why many claim to experience fairies and other ghostly beings exist. They do not know it is they who created them unconsciously. Which this may be the most possible nature why mysticism still remains strong and well accepted by mystical believers.

Just the mention of another parallel world to many scientists is delusional, but maybe it is their own ignorance that makes them delusion? But this is not the case with all scientist, many of those of quantum physics believe there has to be other realms beyond this one. And you will find many different possible realities might look like. Ranging from layered to bubble parallel dimensions. But theories are theories, they are not facts. Facts have to be proven, but it is impossible for the typical minded people see beyond the physical. There is no scientific tool that peek through the many layers, only the human conscious can teleport though these layers. Which only very few common people had experience. This does not mean it isn't possible. But this also doesn't mean what you experience might have been real? Sometimes people create the experience, not teleporting their conscious beyond the body. Which is why we should question our experiences for clarity. We should never accept something as truth unless you truly know the experience was a real teleportation of conscious project. In New Age term "Astral Projection".

There is actually a out-of-body exploration research that is full of personal experience by those claiming to have astral project outside their physical body and tell they experience of what relies beyond this physical dimensions. As to any truth to these experience? We can never just accept nor deny what other people may had experience, do to the fact many seek tell their stories for attention or dream experience that does not mean they actually travel beyond their physical self? But that doesn't mean they didn't either, for these kind of experiences have to be of personal experiences. We just can't go by what others tell us, What you will find is similar description what these other unseen layers of other realms look like and what they encountered in every dimensions. It is more of higher possibility that our existence is layered like an onion, or rings of Saturn. At the core there is a matrix realm, what many would refer to has God and true haven. Outside this matrix realm lies they many layers of existence. From the physical realm to the false heavenly realm, these layers become more and more formless, and are more and more responsive to your conscious thoughts. Which may answer one of the possible theory of a conscious universe? And at the far end lies complete darkness. Like the Yin and Yang, light creates and dark destroys. Therefore both govern existence, they follow the cosmic law of change. And like the cycle of change, existence is never stable, always changing. What was once created one day soon becomes undone. Only to take a new form, new life, new dimensions. This is the best way to describe what these other dimensions look like, and it also explains some of beliefs that many mystics tell their story of what the afterlife looks like, even created their own experience of what they created in the dimensions that create what you image into being? The only way we truly going to know this is by having such experiences for ourselves, Which many skilled meditaters can do, and some people experience during sleep state. Other then that, scientist have no real way of knowing. Theories do not make it a fact, since it would be hard to prove it. Religious people may had some experience in these dimensions, but that doesn't mean that have full clarity on what they actually experienced? What is certain, they do exist but lesser minds can't see what lies beyond the illusion of this physical dimensions.

Chapter Five

Consciousness

Science and religion really have no clarity on the existence of consciousness. For many scientist, consciousness is the byproduct of chemicals in the brain of animal lifeforms. While in many religions, it starts with the gods, than animals and humans. But there has been plant research telling even plants have a conscious too. They are able to communicate with one another, even some lower forms of emotions? Which they show plants show signs of stress, and know when they are in danger. But to many atheists and mystics, this can not be? But facts are facts, and it has been proven. Isn't that what our generation is seeking to know? The truth? Not everybody is closed minded, many of us really want to know this cosmic truth. That is why your reading this book right now isn't it?

Science would like us to accept that conscious is the byproduct of chemicals in the brain, but what are chemicals? Chemicals are made up of atoms, and atoms are the byproducts of energy. So the question here is, does this mean energy it self has a conscious? The best way to answer that is with the question to how does energy know hot to create themselves into atoms and matter? Is it energy the true cosmic conscious that woven this entire cosmic existence into being? If so Then Why? The answer I can give is maybe the cosmos want to existence it self as starts, galaxies, even life forms? Aren't we illusions of nothing more than manifestation of energy? So if we are not "me" as in self but "we" as the make-up of conscious energy particles, that means conscious doesn't existence in the brain alone. The other question you should be asking yourself don't DNA and living cells have a conscious too? How do cells know how does DNA know how to create life? How do cells know how to clone themselves and make up organs and such? How do white blood cells know to attack viruses? How do tissues know how to repair themselves, and so forth? Maybe the answers is conscious exist throughout the body? The brain is just a computer and conscious is the one running everything? Meditate on these for clarity.

Depending on the religion? Now all religions tell conscious started with the gods, and the gods created man. In one tribal belief, it is the cosmos that has a conscious. It is the cosmos that is consciously aware of it's existence. This is what some tribal spirituality tell stories of some cosmic creator and that energy is a conscious spirit. They tell we have two souls, the flesh and the spirit(ghost). The flesh has a mind of it's own, and it is the flesh that dies and not the spirit. The spirit is our conscious, and conscious is energy, and energy never dies. If energy has a conscious, that would mean we are in fact mere illusions that these particles lead us to believe we are "I and me", the ego. Then if that is the case, this would mean there is some cosmic god that is the creator of existence, just not those gods of religion, Well the cosmos created everything, and it dreamed our world and our selves into being, Doesn't mean it is not a god, just isn't a single entity but many particles of couscous. There are quantum physicist that believe our cosmic existence is in fact a live or is one big mind. But beliefs are not facts, and facts are not truth. Truth is the result of the fact being complete proven, which both science and religion are limited to their beliefs rather being open to the possibles of a conscious universe.

What is certain is conscious is not just of animals and human, nor does it only exist in the brain. There are more and more research proven that conscious exist in plants and throughout the body of living beings. Therefore, life is fully away of it's existence. But the origin of consciousness doesn't exist in lifeforms alone. In other words, mind is conscious and conscious is not part of the brain but the matrix. When we die, that doesn't mean our conscious self or spirit fades out of existence. Only the flesh dies, and fades away in decomposing process. But at to our entire cosmic experience being a conscious entity doesn't mean science is wrong of some creator does exist, nor does it mean religion gods do exist. It just means that consciousness is the awareness of consciously existing and perceiving our existence. But conscious is not the byproducts of chemicals but energy particle of many particles manifesting as one consciousness. This may sound very far fetch, but so is religion to science and science to religion. And not all people accept science a lone nor religion a lone. For science only have to go by theory, and are limited to the physical dimension. Religion is full of made-up entities and tell stories in which majority of the teachings do not make any sense. Science can't explain and religion can't explain everything, they are limited to their ego limited understanding. Part of that reasons is do to their own ego beliefs of expectations and their delusions. Scientist must understand logic can not give you the truth, you just have to keep a very open mind and let clarity light up the any open space in which gives us full understanding that is beyond logic. While religious people need to understand religion is out-of-date, and religious minds are not very open to any real truth. If it truth you truly seek, you have to have an open mind and allow clarity come to you. Other words, think less and keep mind empty much as possible.

Chapter Six

Collective Consciousness

It would be hard to accept our world being an illusion while many conscious beings are seeing, hearing, and sensing all that is of this physical world. How can we see the same thing, hear the same thing, smell and taste the same thing, and experience our dimension as something solid and have weight? For many scientist and religious people, just the thought of it not being a solid nor have any real weight is highly delusional. But didn't science knew at a time how the human body really works? Didn't religious people thought of the world not being a spherical shape and rotates around the sun? So how can the thought of our world being an illusion, and what we experience actually is the manifestation of our ego conscious mind? Well at present science is proving that to be the case. It was always known in Buddhism that our reality is a mere illusion, and it only exist because we perceive it. And at present, there are scientist telling how we actually experience our reality together. The best way they can explain it that we are part of a collective conscious, our minds are linked to one another like particles that make up a spider web. This this is not something new, many shamans have been saying the same ting for centuries. Everything we experience of our physical reality we share as linked minds. Like sharing the same dream, which you may not be aware we are all dreaming the something. Science is just coming to be more open to this possibility truth. But does this mean it is a fact? Not if your not aware of it.

At present, there is a new scientific theory that all minds are linked not to one another, but the entire cosmic existence. The most poplar theory is refer to as "entanglement minds" Which explains our minds are linked to one another like one big conscious web of consciousness, and our collective conscious is part of the cosmic conscious existence. Which theory is growing in popular accepts, and help best scientist understand the nature of extra sensory perception, Since all minds are linked, we experience our reality as if we are all in one big dream. This is how to explain why we experience what we sense of our existence and our physical existence. Only in deep meditation of delta state we as ego conscious no long perceives its and our physical existence. Instead, we exist in our conscious matrix or higher self. It is here lies pure consciousness. The all knowing conscious that is linked to the cosmic matrix of our entire cosmic existence.

Many religion would like us to accept that we are individual conscious beings, and that our reality is a real physical entity. But that is not what you dine in the philosophy of some shamanship and Buddhist teachings. Unlike many religion beliefs, much of these two teachings were from experience, not made up. Native American medicine men may tell you that we are all part of the same cosmic web, and that we are all linked to one another. Buddhist tell that all is an illusion, the manifestation of our ego perception. We are not separate beings, we are all linked in this big cosmic illusion. As to if these people actually had such experience are only know to those who experienced it. Even many highly spiritually sensitive people sooner or later experience this cosmic truth in their life time. More so in long periods of meditation rituals. Such has Zen and Raja meditations. This doesn't mean what they experience is what they perceive as truth, For many spiritual people tent to accept the first experience as real, not that they could of dreamed it or an illusion of their delusional state of mind? Which that is one of the biggest mistakes by religious and new agers. For what you experience, doesn't mean you didn't created your own experience. Part to truth seeking is questioning everything, even your own experiences. By experience it over and over can in time give you more clarity what you are experience. A wise spiritual person would just accept the experience as it is, not claiming what it is or isn't. Just as I teaching in my earlier books. It is the same why ghost seekers seek to experience ghosts in haunted places. They really want to know if they truly exist of nor. But you have to be skeptical too of your experience. For believe rs tend to taint their experience with false perception. Which is very common thing.

As to the truth of all minds are linked? It's hard to prove such theory being an actually fact, mainly of those with lesser minds. But for those with higher extra sensory perception, those who have years of spiritual experiences and meditate for a very long time have a much greater experience and clarity of this fact. New agers don't have these higher spiritual experience, because they believe in false and their minds are clouded with their delusions. Highly spiritual ones are truth seekers, not belief seekers. For these higher minded spiritual people, they are very careful how they perceive their experiences, and always questioning that in which they seek clarity on in meditation practice. Belief is not something they seek, religion and new age is not the source of truth. Not even science is. As truth seekers, we can't accept what science and religion belief to be truth. We have to find out for our selves, not accept what others believe. And knowing if we are all connected to one big conscious web of consciousness can only be revel in your matrix conscious. What Raja may refer as higher self, the non-ego conscious. This higher conscious, the matrix that lies under the ego conscious has the real answers. Answers that it want to share. But you have to truly want to seek this truth, and that means you have to take up meditation that allows you to drop from your beta mind to delta. With practice, you can remove your self the the beta mind of logic to the alpha mind that has greater clarity. For it is always in the process in solving everything without logic. And it is the alpha mind state you always want to be in if it's clarity you seek. For clear mind is an open mind, and open minds has greater clarity. It see everything as it is, and knows the nature of things, If its truth you seek, you have to evolve you state of mind to the non logical state of consciousness.

Chapter Seven

Evolution Of Life

Neither science nor religion know actually when and how life came to be on this planet. In science, it began either on land, sea, or both. This is where DNA that manifest into a single cell organism began, and all many other life forms over time. While in such religion like Judaism, God placed both plants and animals before the first humans. When it comes to humans, neither side actually know how man came to be. Atheists are quick to accept we evolved from primates, and in religion humans were just placed here by magic or by gods. In one of my earlier books, Na-Wa doesn't deny that we came from primates. For the fact it tells it was animals that dreamed us into being through millions of years of evolution. This is because Na-Wa is not a religion only natural spirituality same as our plant and animals kin spirits share. No belief, no worship. Just self-exploring spiritual. Which explains the evolution of life in a kind of story that helps you understand evolution with greater clarity.

For a long time, science told us we evolved a fish. Then it said we evolved from apes. Now it tells us we evolved from particles that seem to have a mind of their own in which they made themselves into DNA, which lead to the first life form of a single cell organism. It was from there, all life came from. And it was life forms that changed themselves into new species. It was not from some religious god or extra-terrestrial beings. But now there is a conflict with these theories. For now many scientists are wondering if life even began here on earth, but came to earth from space. And evidence show micro life does in fact exist off world. This tells how scientists claim to have all the answers, just to find out they do not actually know anything. It would just be wise not to claim you know something that you really don't know. In spirituality, we refer to these peoples as fools(not in a rude way). Like an Tao old saying that tells a fool claims to know everything and a wise man does. Besides, wise men don't reveal all, only fools who claim to know all. I don't make any claims, because I want you to find out for your self, and ask a lot of questions to the same knowledge subject you clarity on over and over again until your fully understand it.

Religion on the other hand knows no evolution, everything just existed all at once. And not all religion tell the same story, some are borrowed from other religious beliefs to help fill in the gaps for their religion popularity. In Judaism, God created life after God created the heavens and the earth. It is said God then created Lilith and Adam, Eve came after. You read about earth was made in short time and life came right after that. But we now know that that never really happen. Nor a lot of thing told in the stories even happen. We know race and language never started at Babylon, It all happen during human evolution do to migration. But this doesn't mean certain people and events never happen, because we are now finding out that such cities and events did happen, while other events never did. Although religious people would like lash out saying these are all lies and that we should just believe what the bible tells us. We should never question it. You may find this no different with atheists too. But part of seeking truth is questioning everything. Other wise we may be accepting lies. As many religious and nonreligious people always had throughout history. Even history is full of lies, that just ego nature.

Only in my teachings in Na-Wa you will find evolution of both existence and life on earth told similar to science but with spiritual essence. That is because Na-Wa is not science nor a religion, it tells the story of evolution in a way for you to best understand how the cosmos and nature works. It does not say that it happen in those certain orders, nor claim to know that's how it actually happen. For that fact we were not not there to witness it all. The stories told in my books only give you clarity of the cosmic and natural order. It's no different that you discover fissile that tell us more on the story of evolution, and where certain animals may had evolved from? It just say evidence does not always mean it happen in that order. So Na-Wa teaches never claim to know something if you never witness it as it happen. It also teaches you should deny something all because you never had an certain experience that many other spiritual people have. Know that there is a physical side and a spiritual side of existence. You can't have one without the other. They both go together, like the Yin and Yang.

Chapter Eight

Extra Sensory Perception

Also known to many as ESP or psychic abilities. ESP is the nature of higher sensory perception above normal. Which many scientist find a hard to accepting such possibilities, while there are some that know it does exist. While religion claim it is a gift from gods or spirits. It was also perceived powers of magic or some evil spirit. But fact is that it is not a gift, it is just a higher sensory perception that was mainly used for survival purposes, and to have a non-verbal communication with life and the forces of nature. All living beings have ESP, only the highly spiritual have it greater than those lesser or not so spiritual. Plants have it, animals use it all the time, and we use it mainly to sense danger. It's part of our physical and spiritual instincts, There is nothing magical about it, we all have these capabilities. Common people just have lower sensory perception that those with higher minds. That doesn't mean you can't strengthen it, but for those not born with higher levels of sensory perception and know the ups and downs of it is best left alone.

In Parapsychology, ESP is something that is not fully understood. It was always believed these abilities exist only in the brain, and it is the brain that is studied for proof of the existence of ESP. But that old belief is slowly changing. It seems that the holographic mind theory has a more real explanation of the natural of ESP. It is this theory that mind is not part of the brain, mind existence throughout the entire body and beyond it. You will find this in the "holographic universe" theory as well. But theories are theories, and paranormal researcher seek the facts. Being a spiritual one(psychic sensitive), who like you also was seeker the facts of my heighten senses, as well many of my paranormal experience with no ignorance. I always sought the facts as a kid, and knew I had to find out for my self because I didn't trust science and mystical beliefs. That was part of my self-exploring spirituality, which I came to know of the nature of ESP similar as holographic universe and mind theories had theorize about. Which I knew before I even read about these theories, Science is getting closer to the truth, they just still got a long way to go.

In many religious of the old, we find the beliefs of ESP being gifts from some kind of spirit. This still can be found in modern mysticism known as New Age. Those with these abilities were refer to as oracles which we refer to as psychics today. Not a term I would call my people, but more of those who are fortunetellers. Not all of us accept what mystical people believe, nor misuse our abilities for selfish acts. Those like me only use our ESP for survival purpose, Like sense danger on city streets to unseen beings that are not from this physical realm. Unlike New Age psychics, in my spiritual handbook all actions are govern by the cosmic forces of karma. If you misuse it out of selfish and ignorance, bad things will come. It is not meant to be misused out of egoism. New age also do not have any truth to their teachings. You can't trust what you read, you have to find out the nature of your ESP and paranormal experience though years of meditating on them for clarity. You have explore your self both body and mind to find any truth to what mystic teachings are saying is true. Common sense will tell you many is made-up just to grave your attention to sell books. And I teach these self-exploring spiritual practice for truth and spiritual living in my handbook for those who are actually seeking to know the real from the false, and live a natural spiritual free of beliefs and worships.

If your still confuse, think of ESP as animal instincts. It is the heighten sensory of sight, taste and smell, hearing and sensitivity to external energies that tell a story of your surroundings. Predatory animals use them for hunting, and prey use them to sense danger. When it comes to seeing visions in your mind of things not yet happen, has to do with time is a cosmic window in which you can see all the possible events in which one of them is more likely to happen. That doesn't mean the future is set in stone, it has other possible outcomes. That because the future already happen, but the future is changeable. Time has many windows, it just lesser minds don't know this. But not all visions are of the future, some are just symbolic, which you have to meditate on it to understand what your higher conscious mind or cosmic universe is telling you. And as for ESP being part of the brain is false, ESP belongs to mind, and mind does not exist just in the brain but the entire body. Mind belongs to spirit, your true conscious self. Spirit is only linked to the flesh, and uses ESP to sense the surroundings. But not all of my kind and those of science know this truth, For belief is more easier to accept than digging up the truth that requires years of seeking. Don't let your self be robbed of truth, keep seeking and be patience. You may get it in time, you don't expect science and religion tell you their so-called truth. Truth has to be discover by exploring for it, and meditating on it for greater clarity.

Chapter Nine

Ghosts

Since tribal times to the present day, you will find many beliefs of ghosts in many religions. Before science, the existence of ghosts were as real as the tree outside you home. We find many cultural stores of all kind of spirits, and to many who never encountered these non-living beings find it hard to accept they exist. Since the creation and popularity of science, this belief of ghosts loses it's popularity. This doesn't mean it faded out of the minds of humanity, for the existence of ghosts still strong in the minds of believers and those who claim to have ghostly experiences. For those who never encounter ghosts are more likely be skeptical they exist, but not all are that closed minded. That is because for those like us that seek the truth, they would have to go out and haunt for ghosts at placed that are known to be haunted by many. Just the belief is ghosts doesn't mean they do or do not exist. Knowing is experiencing, and if you experience a ghostly encounter that would be more proof for you. But that does not mean you can make believers out of others, for those non-believers would have to experience these spirits for themselves. There is also the possibility that what you thought you experience was a spirit, might really had been what you protected in you mind that tainted you experience in which you could have seen something that wasn't even there in the first place. We hear about this all the time with people believing they seen a ghost in graveyards, only to find out that it was only a statue on top of a tombstone, or a plastic bag caught in some branches of a bush. Many of these ghostly encounters may also been from the deception of the eyes or delusion of the mind? All because we experience something like a ghost, doesn't mean we actually did. We have to questions these personal experience, and we have to keep experiencing these events to gain more clarity. Other was you accept the false as real. But this doesn't mean ghosts don't exist, because even those most skeptical people sooner or later have a ghostly encounter that defy their understanding. It is then science doesn't seem have all the right answers after all.

For a very long time, science only perceived what they can see and touch. Anything that is not in their line of perception means it doesn't exist. Even the though of atoms were thought to be bull crap. It wasn't until science had evidence with more powerful equipment that proof of atoms exist. It is the same when the notion of string-like energy exist, but until there is some proof it is just some delusional beliefs by certain scientists that in many minds shouldn't even be scientists. It is no different with that of the paranormal. For many scientist, once you die you no longer exist. These people don't know that in death their conscious still exist, but in energy form. But that is not the case with other scientist, for there are those who study the paranormal(parapsychology) that take this stuff seriously. These science people really want to know the nature of ESP, paranormal activities, and make sense of how we still exist after our soul departs from from it's no longer working human vehicle. And there are those who do there own paranormal research calling themselves ghost hunters. But seeking out ghosts is a very dangerous thing, and not all ghosts are harmless. As many ghost hunters can tell you. But not all ghost haunters are very honest people. What you see on television doesn't mean there are some things if not all things are faked. When it comes to the entertainment industry, ratings and sponsors are the big prizes. While many are out to seek wealth and fame, doesn't mean they are true ghost haunters. Doesn't mean they faked evidence themselves, because many want to make believers out of all of us, but that only raises a lot of questions and deplete their purpose. Honesty wins the minds who are open to the possibilities, deception only catches the mind who just want something to believe in. From my own personal experience as what they call "sensitive", ghostly encounters is a daily thing. But I am not out there to prove anything to any way, that just something they have to experience for themselves. And since I have years of experience with spirits, I warn any one who do not know what they getting themselves into that there are in fact spirits that are not friendly and have the power to harm you in a way a strong living person can. Mainly those who are not of my kind, because their light is too dim to protect them. If you are going to do it, purify your mind and emotions with daily meditation and never be negative in mind and emotions. Other wise you may be asking for death wish.

Religion is full of stories about spirits and the afterlife, but not all these spirits were real encounters. It is a known fact that the images of the demons of Catholicism were created to represent once of each of the seven deadly sins. These demons never existeded, doesn't mean a evil departed human soul can't take the form of these demonic creautres. While those of bookieman stories of much earilyer acient religious where not all real enocunters. Most were of dreams and halusination, or just made up as part of their religious politic to populaize their religion and some kind of gain(wealth and power). Then there are stories of ghosts of departed love ones like those in greek culture in which they were feared. To them the sight of departed love one meant the spirit was not happy for some reason and may take revenge or haunt those don't them wrong in some way or another. Which this fear was shared by many in other cultures up to this present day. Although majority of chirstians don't believe in ghosts other than angels and demons, we do find them in jewish folklore. You will find one popular well know evidence straigh in the Old Testament (1 Samuel 28:3–25). In Tibet Buddhism, we find stories of spirits in their book of the dead. With all these stories about ghostly beings, it would be hard to tell what is real and what isn't? That is because those like me rely on intuition and higher conscious to reveal to us the false from the real. While those of mystic beliefs are would just accept them all as real beings, and not that may be made-up or delusion of false perception. In my first book, I reveal what are real and what are not. But I ask you do not accept what I teach, but figure it out for your self.

With all these stories about ghosts, it is no wonder why many are easily ready to deny the existence of spirits while others really want to believe they are real beings. But belief is not a fact, it is more of fiction or delusion. Facts are that in which can be proven or disproved, and that is the real purpose of truth seeking. If you asked me if I believe in ghosts, the answer would be no. No because I am not a believer but a expereincer. And since I have experience with spirit encounters, therefore I know ghost existence. For knowing and believing really aren't the same thing. But all because I know they exist, doesn't mean those who believe or deny know they exist. For to believe in ghost without experiencing them is not knowing, but accepting that in which you have no proof they do or do not exist. Same rules apply to atheists. As to how is it possible for our conscious to exist long after death? Really there is no real answer to that question except it is what it is, and that is just how it is. As to why spirits are not always in their human form, not the same age as they were when they departed, even shape-shift into some other being? First you have to understand that your spirit isn't a single entity, it is made up of countless particles of energy that have a conscious. Together, they give you the illusion your a single conscious being. Your no more than one big cosmic hologram, a dreamed up being by the cosmos. And energy does not have a sex nor form. Therefore your spirit really is sexless and formless. You are how you perceive your self, and that is why ghost look like or are way different that you expect. In my spiritual teachings of Na-Wa, you will find that spirit true form is orb shape. That is because departed soul lose their attachment to their human form, but still hold on to their ego conscious. The color of the orbs is the auric color of their spiritual state. Green to white are safer than those yellow to red. That is because green to violet are spiritual lights, which tell us how friendly they are. Those of yellow to red tell us we need to be very cautions of these spirits. Mainly those of red orbs, that is the hot aggressive ego that is of evil nature. In other words, you demonic ego self. While the violet color is your highest spiritual ego conscious self. The white is the non-ego self, pure conscious.

Chapter Ten

Cosmic Conscious

Since the beginning of humanity, ancient people had a feeling of a presence that govern our universe. Despite what atheist want us to believe, the belief is a cosmic god wasn't always something told in religious stories. For those of tribal past, they felt there was a great creator presence among them and the planet. But at that time, God had no form just sense of a presence, even a divine experience. This was the reasons what spiritual rituals were part of their lives. Rituals that can still be found in many of the old and modern religious today. It wasn't until humans started to migrate and settle in cities they build is when God started taking on forms. But many of these gods had political purposes more those have spiritual ones. And it is these different beliefs that made accept any religion to be something real. Which is the reasons why science is the well most accepted that explains everything that religion can not. That doesn't mean science has all the real answers, science is more of study and facts. Religion is more of story telling and beliefs that does not make any sense for those who have a much more understand of things. But that doesn't mean there isn't some cosmic conscious that governs the entire cosmic existence. But it may not be the God of what you want to accept from popular religions like the God of Judaism that is most well known spirit to many. The God of the cosmos may not even be a single entity but manifestations of other energy particles that make up this one cosmic conscious that religions refer to as god or gods?

Atheists are quick to deny the possibility of some kind of cosmic conscious truly exist. Just the thought of God brings out their inner demon self. But yet they treat science is if itself is a god? This is one of the reasons why atheism is seen a a religion. They believe what ever science tells them, and like an occult many of these science worshipers play god. But this kind of ego self worship has karma govern over it. You may heard the old saying telling you do not want to mess with Mother Nature? That is what the saying means, not to play god with that is govern by the natural forces in which can destroy you for breaking the laws of the natural order. Breaking these laws do have serious consequences, just atheist don't know karma is actually a real force that governs all actions of life forms. It is not God that does the punishing, its their own karmic actions that will do the punishing. This goes for all people both religious and non-religious. You can't escape it, all your actions has a karmic effect. That's just the natural law of the cosmos and of Nature.

Religion on the other hand focus on some higher power, and create gods that serve their religious purpose. If you look in Hinduism, you find each gods represent something. We also find this in Greek gods as well. You only find only very few religions that have only one god. While the Romans and Christians had worship a living persons in the belief they are god descendents. But the cosmic law prevent humans being gods, although many religious people refuse to accept this cosmic truth. But even common people play god, and expect to be worship. We see this everywhere, that is the nature of the human ego. Why else there is war, class of people, worship of departed people, and so forth? Its the ego that plays god, it is not the higher conscious that has compassion of all living beings and the environment. But no mater what these religions believe, they all are worshiping the same God in which their egos made gods in their own image. Yes, your evil ego has horns, and you higher conscious has angel wings. A real cosmic conscious that many refer to has God has no ego, therefore it has no real form, any sex, nor ego desires. That is only of the human ego, not the real conscious of the cosmos. That doesn't mean that the universe isn't conscious-less, other wise it doesn't explain how prayers are much of the time are answered. Doesn't explain why the cosmos seems to know what is on your mind, and brings that to light in your awaken experience. Like when you think of a televisions show you haven't seen in very long time all of suddenly popped up on television. Some kind of cosmic conscious is responding to your thoughts and urgent needs? Doesn't mean it will respond to negative pleas and egotistic desires. The cosmos is why smarter than you like to believe, it is all knowing and picky of what you pray or wish for. Funny of all believe the universe isn't consciously aware of it self and us.

Here are the facts, gods of religions are not the true conscious conscious. Our entire cosmic existence is one big cosmic conscious that are made up of tiny particles of other cosmic conscious that are linked to form the one. Like snow flakes that make up a snowball, that's how many particles of energy make up one endless cosmic conscious. Know this is one big cosmic dream, and you are part of the illusion. It is the cosmos that is experiencing it self as you and me, all the life forms and planets, stars and gasses, and so forth. It created us, and made us believe we are who we thing we are, and not what we really are. That is one big cosmic holographic image that is no different from creating movie character created on high-tech computers that is in the movies now-of-days. It is also our ego that creates it's own heaven and hell, which can manifest as the place of rebirth in which your karmic actions rewards or punish you to. Unless you are speared by not making enough bad karma that you might be reborn in flesh again for another chance in life. It would be wise to practice being good and doing good for others. As for the truth of a real God exist is something you have to experience for yourself. As I taught in my first book "The Psychic Wisdom Seeker Handbook", to know God exist you have to really experience God, Otherwise you really will never know. Belief isn't truth, but experience can give it as long you have highest clarity that the experience was the real deal and not your imagination. It is the same as the universe having a conscious. You can't really accept or deny it unless you try to experience for your self. Do this through out your life, you may find that there is? Well you won't know until you seek it out. Until then, do not believe but seek the truth for your self. Be your own investigator, your own explorer for the truth.

Chapter Eleven

Cosmic Perceptions

What you see, hear, and sense is just a mere illusion in which what quantum physics and Buddhism tells us. For all that perceive is what our ego perception creates, and what it creates is experienced as something realm In some shaman stories, they tell really is like a dream. It only seems real, but it's an illusion. Therefore we perceive our own experiences, and we share an collective experience we call daily life. What is real to you, might just be something your ego conscious is projecting? What we perceive as ourselves being something real, something living, is all an illusion of this big cosmic dream of perception. Therefore all scientific beliefs of mater and religious beliefs of deities and supernatural powers are all an illusion that is all manifestations of the delusional minds of humanity. It's what you can't see and sense is the cosmic truth, and that truth is not what these believers want to accept. Funny how people claim to seek the facts, but not the actually truth. But believers seek only belief, and truth is very painful to those only seeking and protecting their beliefs.

For a long time, old physics always accepted that everything is made up of matter. There can not be something else beyond that. Then came the belief of atoms, which wasn't well proven until science had the right equipment to perceive it. But there is a problem with this realness, for there was research had been recorded that when looking at an atom something unexplainable happens. When observing the atoms, they seem to fade in and out of existence when the observer is looking at them. But many physicists are quick to ignore this, which their only answer is this is what these observes are just imagining. But this research is being taken more seriously at present, for this supports many of quantum theories of our reality is one big illusion. Other worlds matter is not what old physics thought it to be, and that our universe seems to have a conscious that allows us to see things that are not really there. It's as if the cosmos is the great magician, the dreamer of dreams. Maybe that is the case? But lesser minds will never have no real way of knowing because there are just things can't be experienced beyond the delusion of the ego deception of perception. It is only when you weaken your ego that the illusion starts to fade. The mind becomes the alpha mind, and the alpha mind is not easily deceived. And there are many higher minds that have this alpha mind state. This is why they have much greater experiences of the real reality that common people may never experience.

Believers of the supernatural are also so easily deceived. They lack the clarity in which could help them understand many of their paranormal experiences are not always what it seems? All because they see something in the image of an image of their belief systems, doesn't mean they were actually there. It is a known fact that we can project our false experience, for our mind is also a image projector. It can give us false experience of seeing deities, ghosts, even extra-terrestrials. But that is not saying some of these experiences aren't real, it just saying what you experience might not be what you want to accept. Being a sensitive, it is very easy fopr my kind to see spirits, but not all we experience were real entities. You can easily project your own ghostly experience, just as you can mistaken seeing spirits do to the fact the eyes give off energy that give the illusion something was seen at the comer of your eyes. And I telling this from personal experience in which I don't take the experience as something real unless I actually know what I experience is the real deal. I know my eyes play tricks on me, and I can see things that are not there do to lack of sleep in which your mind project things that are not really there. Only a keen mind has greater clarity, in which is always skeptical what it may had experienced to be something paranormal. Having this state of mind, I can easily say I know what I experienced is either the real or something that wasn't. But much of the times I am never fully sure of many of my experiences, and remain skeptical until I have more clarity in which I have to experience something over and over again and again until I have full understanding what I was experiencing. But many of my kind don't have this kind of mind state, because they are so closed to any real truth that threatens their beliefs that they just don't want to have this mind state. This is actually the same with religious people, but also scientists. What is certain is perception is what it is, a projection of false manifestations that energy creates as something realm. It is perception that is the greatest deceiver of our ego minds that work with the deception of the cosmos. Pretty much like a dream of dreams, and our experiences is just one big holographic illusion we call perception.

As you can see, our experiences is all the projection of both the cosmic and ego perception. Think of the cosmos as the dreamer, and the energy particles are participating to create that in which the cosmos is imagining? Those particles of energy manifest everything that can be perceived. They manifest themselves are you ego self, and it is your ego self that projects your experiences. In other words, your dreaming even though your awake and not sleeping. What you see, hear, smell, taste, and feeling are illusions of your ego perception. Perception has two sides, the false and the real. It is only in the middle you find the truth, and that truth is noting is really real nor really false. It is just the nature of perception, and it is perception that creates our entire human experience and daily life experience. So no one side of the human ego beliefs are right nor wrong. We all perceive differently than others, and that means no one experiences reality the same as someone else. What one scientist perceives as real, the other may not perceive the same thing. One religious experience will not be the same with some one else with a different belief and have a different experience from that other religious person. No matter what the experience is, is all a deception that is both real and not real. Truth hurts, but isn't this is what many are seeking?

Chapter Twelve

Difference Between Spirituality And Religion

One of the misunderstood fact by both atheists and religious people is the lack of clarity of knowing the difference between spirituality and religion. Spirituality is the essence of the soul and living as a free-spirit person who is fool of compassion and shares that compassion in many ways. That doesn't make it a religion, for religion is the belief of the supernatural and of worship of an entity of a living and non-living beings. Such as the belief in god and goddess, worship a person or spirit with great powers, and the belief of false truth in which bother atheism and religion share. Therefore spirituality is natural, and being spiritual just means being good and doing good for fellow man kind, other living beings, and taking care of our environment and wild habitats.

Since can be seen as a religion, for the fact many scientist believe in things that they have no real proof of, and things that can't be explain with logic so therefore it isn't something real. As explained before, the belief in the Big Bang does to the perception of seeing our universe expanding doesn't mean it happen. Nor doesn't mean the universes is really expanding out of existence from human observation, but a mere cosmic illusion in which defy logic in which you can't really explain. Although many of us spiritual ones do not deny humans evolved from primates, but there is a lot of empty gaps in this belief that many atheist do not have any real answers to explain a lot of things. Such as if we evolved from primates, why are there still primates existing today? The answers you can only give is that it is highly possible of branching of species and mutations. So we get a different kind of primate that was our ancestors. But there is another problem, and that is how did humanity became so intelligent to build cities, use math, even build things beyond our evolution nature? Problem is atheist don't have any real answers for this conflict issues. But there are some that would quickly say we were of extra-terrestrial experiment. That would be the best explanation for this issues, but there is still the matter of having the real facts. How would we really know this to be the case, if we have no real facts to back this up? This is were we find belief in science and of atheism. No matter what these people want to accept, they are no different from religion.

Religion is full of things that make no sense to logical minds, no holds very little truths. For if religion was true, then why is there more than one religion that believes something totally different? If religion was true, why are there things that don't explain things that we discovered that religion does not mentions, nor had the capabilities to explore for? Such as the first life form, dinosaurs, evolving from primates, the earth rotates around the sun, why we do not see these gods in person, and so forth. The only real answer is because they really do not know the truth, nor have all the facts. Just like science. It is easier to fill in the gaps by making up stories that best explains everything, and magical events that draw people into believer in these religions. The questions we should be asking is did any one witness these events? Were their any written records to back them up? Did a man really walk on water or lift is arm and flew up into the sky, or was that added later to draw people in be more acceptance to religious popularity and it's political control? No, people just accept without questing if these events happen or just fiction. But no matter what people believe, religion can't give you much truth. Nor can it make you spiritual and closer to the divine. To be spiritual means being more in tune with your higher conscious, and being and living a positive life that harms no living being nor the planet. Being spiritual simply means to be good and do good things. Being spiritual means your a loving and caring soul that want to share your love and compassion with other people, other living beings, and caring for our environment. And only you and you a lone can make you this spiritual person, religion can't. Only you can change you life for the better, and make other lives better by giving up you time to help those who are in need. Like helping take out your neighbors trash when he or she is not able to do so physically, give a dollar or so to an homeless person out of compassion, put a smile on someones face when they are sad. This is what being spiritual is, and you charity not just brings joy to others by rewards you with the good feeling with joy from doing so. You become happier when you make others happier. That's is what being spiritual is all about.

Therefore spirituality is the essence of existing, and being spiritual is the essence found in your pure conscious or matrix conscious. All plants and animals are spiritual beings, and they live a natural spiritual life. They know no religious gods, worship any beings nor forces, nor are evil nor holy. They just know their existence s and the existence of other living beings and their surroundings. They live in harmony with other living beings, and enjoy the company of them. We find the same qualities in Native American and other tribal spirituality. You will also find it in my spirituality I just refer to As Na-Wa, which is natural spirituality. So you don't need religion to be spiritual and live a spiritual life, and living a natural spiritual life will give you more peace and more joy in your life. You can think of natural spiritual as "Nature is my religion, living in peace and harmony is my spiritual practice"? You may find meditation is a part of natural spirituality. For all living beings meditate naturally. The tree meditates on the sun rays, the bees meditate on finding nectar, the rabbit meditates in a nice cool breeze, the fox meditates on observing a mice its going to catch for food, and bear sleeping in a cave, and the dog meditating on the sound that gets its attention out of security reasons. So do you meditate naturally. You meditate by focusing on something, meditate on sensing something that needs your full attention, even sleep is a meditation. So just because people practice meditation for inner peace and make their life more enjoyable, doesn't mean it's a religious practices. It is part of our spiritual essence in which we do unconsciously if nor consciously. That is why I typically refer to my spirituality to those who need more clarity as "Nature is my religion, and meditation is my spiritual practice". Even thought it is not a religion, others see it that way because they lack clarity of the difference between spirituality and religion. And being spiritual just means being positive in mind and being, and being compassionate towards a living beings and care for our planet's health. Your love is a spiritual quality, being unloving isn't. You caring for a love one is a spiritual quality, not caring isn't. You treating your animals friends like your own children or best friends is also the qualities of your spiritual nature. If you did not of those spiritual tings, then you would be considers nonspiritual even wicked. That's the difference between being spiritual and being religious. Therefore you do not need to be religious to be spiritual, you just need to be spiritual and life a natural spiritual life that will make your life more pleasant.

Afterwords

I like to thank you for taking the time to check out this book and taking the time to read it. Please accuse any bad grammar, because I have dyslexia so that will be expected from me. Please show some respect, and be open to my teachings. I can only hope it help open your mind to such possibilities, and that it insure you to be your own truth seeker. Know that all because science do not accept a lot of things of the supernatural being real, doesn't mean it isn't. Only experience will give you the real answers. As for seeking out ghosts, I do recommend you be open to living a natural spiritual life that can help raise your inner light that can protect you. You may want to read my first book on living a natural spiritual life, which is the spirituality of "oneness" and living in harmony with all that we share this planet with. In other words, nature is your spirituality and meditation is your spiritual practice. No belief, no worship is present in this book. It has more information about spirits, and how to defend your self from the most evil ones,. As well learn how to use your own ESP to detect spirits, mainly those that can do you extreme harm. I advise not to misuse your spiritual perception for selfish gain and wicked purposes. Other wise you may not just create bad karma for your self in which you will regret, but it also attracts evil spirits waiting to do you harm. Which that would be result of you bad karma punishing you. And know what you do not need religion to be spiritual, you just have to be a good person and be and do good onto others. That is what being spiritual is, and it brings out the best in you and your present can bring joy into the lives of others. Other words, be pure in mind and pure in being. Your light will grow and protect you and attract positive people in your life. Don't just take my word, just find out for your self with an open mind and full awareness that will give you clarity.